Quantum Physics and Quantum Mechanics for Beginners

The Introduction Guide For Beginners Who Flunked Math And Science In Plain Simple English

Donald B. Grey

Bluesource And Friends

This book is brought to you by Bluesource And Friends, a happy book publishing company.

Our motto is **"Happiness Within Pages"**

We promise to deliver amazing value to readers with our books.

We also appreciate honest book reviews from our readers.

Connect with us on our Facebook page www.facebook.com/bluesourceandfriends and stay tuned to our latest book promotions and free giveaways.

Introduction

If you are interested in understanding quantum physics, but are turned off by its latent complexity—the mathematics and formulas—you are in the right place. This book was written to make quantum physics clear and understandable for those who find the study of math and science challenging, to say the least. You have every right to revel in the amazing discoveries that define and explain the subatomic world within our vast, expanding universe and everything it contains.

What is the difference between quantum physics and quantum mechanics? They mean the same, so we'll use the terms interchangeably as the science of matter and energy on the subatomic scale. According to Stephen Hoyers, Ph.D. in Physics, UC Berkeley:

> "As most physicists use them, **there is no difference** between quantum mechanics, quantum physics, and quantum theory."

From the earliest days of quantum theory more than a century ago, which began with Max Planck and Albert Einstein, through the present, quantum physics has been praised as brilliant, breakthrough science, and criticized as illogical, incomprehensible, and in violation of the known and trusted laws of classical physics. Yet increasingly sophisticated experiments and observations are confirming theories and hypotheses. Quantum mechanics is a reality.

Science still does not yet fully understand the complete nature of matter and energy, which has triggered vigorous and spirited debates, theories and experiments, which we'll review in the upcoming chapters. The quest for a grand unified theory, or "theory of everything," that unites the forces of the universe, continues. Let's begin at the beginning for an overview of this amazing universe that we are a part of.

In the Beginning

Our understanding of quantum physics begins with the formation of the subatomic particles—the quarks, electrons, protons, neutrons, neutrinos—and the atoms of which they are components of. The word "atom" means "indivisible" in ancient Greek, a point at which matter can no longer be divided. For over 2,000 years, this definition, in its simplicity, was as accurate as science could determine. But, during the 20th century, theories became verified by experimentation, and quantum physics came into being.

The Big Bang. Approximately 13.8 billion years ago, the universe was born in a phenomenal outpouring of energy—an event commonly called the *Big Bang*. Scientists have tracked backwards to within trillionths of a second of that expansion, which continues today, with no foreseeable end in sight. All that exists today in the universe—from subatomic particles, atoms, elements and molecules, to trillions of stars and planets, massive galaxies, black holes and quasars at the far edges of the universe—are composed of matter and energy created in that initial moment.

The Big Bang, 13.8 billion years ago

Inflationary Theory. In less than a billionth-trillionth-trillionth of a second, the universe went from a virtual point or singularity, far smaller than a subatomic particle, to the size of a billiard ball, a hypothetical phase called "inflation". This was the only time that matter and energy traveled faster than the speed of light. While inflation has not been definitely confirmed, it is increasingly becoming accepted within the scientific community.

Cosmic Microwave Background. In those first seconds after inflation, virtually all the energy of the expanding universe was in the form of heat, with temperatures reaching 10 billion degrees Fahrenheit (5.5 billion degrees Celcius); it would take 378,000 years for the universe to cool sufficiently so that the subatomic particles and forces can combine and form complete atoms. This caused light to appear for the first time, in a burst of light, across the universe observed today as the Cosmic Microwave Background, or CMB.

Enter the Quantum World

The following chapters will take you from the early days when quantum physics was first conceptualized, to the evolution of theories, testing and confirmation through experimentation, and on to the latest 21st century ideas and discoveries. Here are the highlights of each chapter:

Planck and Einstein. You'll meet the physicists who started the quantum revolution. The most influential was Max Planck, acknowledged as the father of quantum physics. It was Planck who first theorized that atoms received energy and radiated energy in quanta, or finite bundles—the smallest possible units of energy. Planck was among the first to recognize and support Einstein's Special Theory of Relativity.

Wave-Particle Duality. The concept, in which light can travel as either a wave or as a particle, was derived from Planck's quantum theories, and published by Einstein in 1905 as the photoelectric effect: The concept that light travels in quanta, or discrete bundles of energy, called photons, became accepted science through tests, including the double-slit experiment, and the Compton Effect.

Subatomic Particles and Forces. An atom is a nucleus composed of protons (except for hydrogen) and neutrons, and encircled by electrons; one negatively charged electron for each positively charged proton, resulting in a neutral charge. Within each proton and neutron are three quarks which affect the electrical charge. Three forces within the atom—electromagnetic, strong, and weak—mediate the attraction that holds atoms together to overcome repulsion, and cause conversions and decay of the particles.

Neutrinos and the Dark Sector. After the initial 5 to 6 billion years of slowing, the expansion of the universe began to accelerate; today, it is attributed to a mysterious force, called Dark Energy. Another mysterious entity, Dark Matter, is asserting a powerful gravitational force that holds stars and galaxies in their tightly constrained orbits. Are ghost-like neutrinos implicated?

Bohr-Einstein Debates. Albert Einstein and Niels Bohr debated concepts, hypotheses, ideas, theories and challenges to our current scientific understanding of the rapidly-evolving discovery of the atomic and subatomic world. Are light and electrons waves or particles? Their debates were high points of scientific thinking early in the twentieth century. Physicists concur that Bohr was successful in his defense of quantum theory.

String Theory. Also called "superstrings", this concept describes the smallest particles as tiny string-shaped entities, instead of more traditional concepts of subatomic point particles with zero dimensions. The strings are believed to vibrate relative to the charge and mass of the particles they form. String theory was proposed in the 1980s to unify the four known forces in a single quantum perspective—the long-sought *unified field theory.*

Schrödinger's Cat. In 1935, Edwin Schrödinger famously created a thought experiment to demonstrate duality in quantum physics: If a cat and a radioactive element were in a sealed box, you could not determine if the cat was dead or alive until you unsealed the box, implying that until the box was opened, the radioactive element *did or did not decay*, and the cat was both *alive and dead* at the same time.

Entanglement and Quantum Computing. Can elementary particles become linked over a distance, so that when something affects one particle, it also affects the other particle, regardless of how far apart the particles happen to be? Einstein called it "fuzzy action at a distance." This unusual behavior of particles linking or interacting over a distance defies current theories, including limitations imposed by the speed of light, yet experiments are encouraging.

Parallel Universes. Is the universe that we know of actually one of many universes, floating in a network of bubble universes? Beyond the imaginations of science fiction writers, astronomical theories are exploring concepts of a multiverse or parallel universes: Vastly diverse universes, or universes that are virtually identical.

Okay, with a promise of an enlightening, math-free adventure in quantum physics, let's get started.

Donald B. Grey

Chapter 1: Planck and Einstein

Albert Einstein modestly stated that he "stood on the shoulders of giants."

Is the name Max Planck familiar to you? Anyone who studies physics knows him as the father of quantum physics, and as one of the most influential scientists in history. He is credited with the concepts that form the basis of quantum physics and mechanics, and was the principle "giant" upon whose shoulders Albert Einstein claimed to have stood.

To put his importance into perspective, today, there are 86 Max Planck institutes and research institutes worldwide, with most in Germany, and all operated by the Max Planck Society. These facilities are divided into three research areas: Physics and technology, chemistry, biology and medicine.

So who exactly was Max Planck, and why is he credited with being the founder of quantum physics?

Max Planck

It was Planck who first theorized that atoms received energy and radiated energy in quanta, or finite bundles, the smallest possible units of energy. His discovery, for which was awarded the Nobel Prize in Physics in 1918, is regarded as the birth of quantum physics. Planck was among the first to recognize Einstein's Special Theory of Relativity in 1905, and his influence in theoretical physics ensured that the theory was widely accepted.

Planck was born in Kiel, northern Germany, in 1858, with the full name of Karl Ernst Ludwig Marx Planck; he later adopted the nickname Max, which became his first name. His background was intellectual, with his father being a professor of law, and his grandfather and great-grandfather having been professors of theology. Max Planck and his family moved to Munich in 1867, where fortuitously, he was tutored by an eminent professor, Hermann Mueller, who awakened Planck's interest in mathematics, physics and other sciences, including astronomy.

In 1874, the young Planck was also deeply interested in music, but at this pivotal time, he chose to focus on the study of physics—theoretical physics in particular. As his studies and career continued, he became a professor at the University of Munich and later, the universities of Kiel and Berlin. In 1894, he began to distinguish himself with new theories about black body radiation, which is concerned with how energy is radiated or emitted from a perfect absorber. His concepts challenged traditional theories, and thus began Planck's deeper investigations into how energy travels.

The Planck Constant

His breakthrough came five years later in 1899, with his conceptualization of light having discrete, or finite, dimensions. The energy of light had previously been assumed to have a continuous flow, like fluid; Planck determined that photons of light have energy in packets, or quanta.

> Let's pause for a second and clarify what a *photon* is. Light is a form of electromagnetic energy, along with electricity, radio, X-rays, microwaves, and magnetism, that can travel as either a wave or a particle; the 'wave-particle duality' we'll get into in a later chapter. The term "photon" describes a particle of electromagnetic energy.

Varying amounts of energy are multiples of a whole integer, or a constant, now called the "Planck constant". Light and all other waveforms are emitted in multiples of the Planck constant. Said another way, the energy of a photon can be determined by multiplying the frequency of the photon (how many times it vibrates per second) times the Planck constant, which is represented by h in formulas. I promised no math in this book, but for the record,

Energy equals the Planck constant times frequency ($e = h \times f$).

Planck went on to designate Planck length, Planck energy, Planck time, and Planck temperature: Each are infinitesimal small units that simplify measurements for use in formulas.

Max Planck knew he had broken through constraints in understanding the quantification of light and other electromagnetic radiation, but he recognized that further interpretation and application was necessary; this would come six years later, in 1905, when his colleague Albert Einstein would apply Planck's quantum mechanics to his theory of Special Relativity. But Planck's work was regarded as an intellectual breakthrough, which is why this work earned him the 1918 Nobel Prize in Physics.

Special Relativity

Einstein's now-celebrated publications in 1905 included his Special Theory of Relativity, and while today it is considered one of the greatest conceptual accomplishments, at the time of its publication, few understood or appreciated what Einstein had accomplished. Fortunately, Max Planck—whose work underpinned Einstein's—not only understood Special Relativity, but was committed to helping it gain recognition, and further, to add his own contributions to the theory.

From 1905 through 1909, when Planck was president of the new German Physical Society, he used his considerable influence to promote Einstein's work. Then, after Planck was appointed to become dean of Berlin University, he invited Einstein to join him in Berlin, where Einstein was awarded a professorship, enabling them to continue and expand their collaboration, which continued for many years.

So, what exactly is the theory of Special Relativity, and why was it so hard for physicists of the time to understand it?

Special Relativity unites space and time for an object that is traveling in a straight line, and is especially relevant to things traveling at, or close to, the speed of light, which is 186,000 miles (300,000 km) per second. The theory postulates that as the speed of an object increases, the mass of the object also increases. While the increase in mass is virtually unmeasurable at lower speeds, as the object approaches the speed of light, its mass increases dramatically to be almost infinite. Special Relativity says that the object, or any matter, cannot go faster than the speed of light.

Einstein used the example of a person in motion, relative to a person who is stationary, and their relative perceptions. It is Einstein's famous mind experiment involving a moving train.

Imagine a train being equally between two trees situated along the tracks, one tree ahead, the other tree behind where the person on the moving train is sitting. The other person is standing on the platform, directly opposite the person sitting on the train as it goes by. At that very moment, lightning bolts strike both trees simultaneously. According to Special Relativity, the person on the train would see the bolt hit the forward tree (that the train is approaching) before hitting the other tree (which is receding). The observer on the platform would see both trees struck simultaneously.

Adding to the applications of Special Relativity, a few months later in 1905, Einstein established the relationship of mass to energy; proving they are one and the same. You know what's coming. The most famous formula in history: $E = mc^2$, or energy = mass times the speed of light squared. In simplest terms, a small amount of mass can be converted into a tremendous amount of energy. Really? Yes, it's the basis for nuclear energy.

$$E = mc^2$$

Chapter 2: Wave-Particle Duality

Is light a wave? Physicists long believed that if light were a wave, it would need a medium to travel or propagate through the vacuum of space. But if light does not travel as a wave, what else could it be? So, they thought, there must be a medium in space to carry waves.

Centuries ago, early physicists like Christiaan Huygens and Isaac Newton debated the form that light takes to travel, or propagate. There had been controversy over the apparent contradiction that light might take alternative forms, depending on the medium in which it is traveling. Certainly light can be in the form of waves as it travels through the atmosphere, which has substance, but what about outer space, which is a vacuum?

For many years, it was believed that space contained a substance called *ether*, which enables light waves to travel. But then experiments conducted by Michelson and Morley (who had distinguished themselves by accurately measuring the speed of light), proved that ether did not exist, and the concept of light being a particle gained support. What was missing at this moment was an explanation of how light could be *both a wave*, to travel through air, *and a particle*, to travel through the vacuum of space. As Albert Einstein stated:

> "It seems as though we must use sometimes one theory and sometimes the other, while at times we may use either. We are faced with a new kind of difficulty. We have two contradictory pictures of reality; separately, neither of them fully explains the phenomena of light, but together they do." *(Einstein, A. 1938)*

This chapter will clarify these mysteries by explaining the photoelectric effect, the double-slit experiment, and the Compton effect, which verified the wave-particle duality.

The Photoelectric Effect

The photoelectric effect is the release, or emission, of electrons from a metal surface when struck by a form of electromagnetic radiation, like light. The released electrons are described as photoelectrons. Prior to

1905, classical physics attributed this emission of electrons to the simple transfer of energy from the incoming light to the electrons. The greater the intensity of the light, the greater the number of electrons that would be emitted.

Or so it was thought. But then experiments failed to confirm this effect, and instead, tests demonstrated that the electrons would not be dislodged or set free until a threshold frequency was achieved. If the frequency was too low, it does not matter if the light is intense, or if it arrives on the metal over a period of time. Why? Einstein believed that if light arrived as a continuous, flowing wave, the threshold needed for electron emission could be achieved given sufficient time, which, as we've just noticed, is *not the case*. Instead, in 1905, Einstein said that light traveled in a series of discrete wave packets, or photons, building on Philippe Lenard's discovery in 1902 that the frequency of light, not its intensity, was responsible for the release of high-energy electrons.

The Double-Slit Experiment

As was mentioned, as the 20th century was starting, physicists like Max Planck and Albert Einstein challenged the assumption that light was exclusively a wave, and Einstein had borrowed from Planck to suggest that light traveled in discrete, or finite packets of energy, quanta, a single unit of energy.

Earlier, in the 1800s, a British experimenter, Thomas Young, had set out to prove that light was a wave. He sliced two vertical cuts into a sheet of metal, and directed a light through the two slits, and on to a screen. The result, as he expected, showed a pattern of bright lines where waves overlapped, crest-to-crest, and dark lines where wave crests-and-troughs overlapped and cancelled each other out. The combination of bright and dark lines is called an "interference pattern", and, at that time, confirmed that light was a wave.

But when Einstein and others later applied detection devices to each slit to track which photons passed through them, the pattern on the screen changed completely, from an interference pattern, to two narrow slits of bright light, indicating that light was now passing through as particles.

What's going on here? This is an example of how quantum mechanics can be weird, and contrary to what we think we know. The very act of observing the photons of light caused them to assume a single position. This will come up again when we discuss Schrödinger's Cat and Superposition.

The Compton Effect

In 1923, in an effort to further prove the particle nature of light, Arthur Holly Compton conducted experiments that earned him the Nobel Prize in Physics in 1927. Compton fired high-energy photons, especially X-rays, towards a target substance, whose electrons in the outer electron shell were loosely bound, and of low energy.

The result, called "Compton scattering", resulted in the loss of energy by the photons, and a gain of energy by the electrons, proving that the photons were particles, because waves could not have the "punch" to affect the wavelength shift that occurred.

Here are some additional points to help clarify the phenomenon of the Compton Effect:

- X-rays were used as the source of electromagnetic energy because they are of very high frequency; therefore, high energy.

- Electromagnetic energy covers a wide spectrum, from high frequency, high-energy microwaves and X-rays, through visible light, and all the way to very low-frequency radio waves.

- The term "photon" was coined by an American chemist, Gilbert Lewis, to define a "quanta", or bundle of electromagnetic energy.

- A photon contains both momentum and energy, in common with particles of matter; and, a photon has wavelengths and frequency, which are characteristic of waves.

- In Compton scattering, photons collide with low-energy electrons and transfer some of their energy and momentum to the electrons, which then become higher in energy levels. The incoming photon then departs at a lower energy level.

Photons of light as particles and waves

Chapter 3: Subatomic Particles and Forces

We'll begin with the basics.

The **atom**, as you know, is the fundamental building block of the universe, and is a nucleus surrounded by electrons. Every element—starting with hydrogen, the lightest, through uranium, the heaviest naturally occurring element, and continuing through all of the heavier man-made elements—is made of atoms. Atoms combine to form molecules, but at their base, every molecule is made from atoms.

Subatomic Fermions: Protons, Neutrons, Electrons

The nucleus of every atom contains protons, and every atom, except hydrogen, contains neutrons (although certain forms of hydrogen, called "isotopes", can contain one or two neutrons). Protons and neutrons are classified as "fermions", a category of subatomic particles with mass.

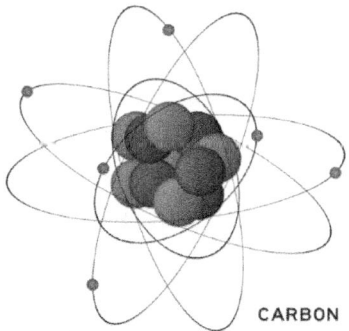

CARBON

The planetary model of the atom. One electron for each proton. In reality, the electrons can be either particles or waves, located a great distance from the nucleus.

Generally, the atom carries a neutral electromagnetic charge, although its nucleus is composed of positively-charged protons, and neutrons, which carry no charge. So how can an atom have a neutral charge; shouldn't the protons create a positive charge?

Well, yes, the protons are positively charged, so to balance their positive charge, we need to look outside the nucleus to the atom's surrounding cloud of electrons that are negatively charged, with exactly the opposite charge of the protons. Because there is exactly one electron for each proton, the positives and negative cancel each other out, and the overall atom has a neutral charge.

Electrons can be added or subtracted under certain conditions, resulting in an ion, or a charged atom. An ion that has lost electrons has a positive charge, and vice-versa.

Atoms are popularly depicted in a solar system design, with a central nucleus closely surrounded by electrons. In reality, there is considerable space between the nucleus and the surrounding cloud of electrons; an analogy suggests that if the nucleus were the size of a marble in the center of a stadium, the electrons would be orbiting just outside the stadium.

Inside the nucleus. Let's take apart the nucleus of an atom and see what's inside the protons and neutrons. We're able to do this because atoms have been smashed together at speeds approaching the speed of light, so their components can be released, tracked, measured and studied. This has been done with increasingly larger and more powerful particle accelerators; you have probably heard of CERN, on the French-Swiss border, with its 40 km accelerating ring.

Subatomic Fermions: Quarks

Looking inside protons and neutrons to see why they have positive or neutral charges, we discover the presence of subatomic particles called **quarks**, which were first conceptualized in 1963 by a visionary physicist, Murray Gell-Mann, and later confirmed by particle accelerator experiments. Quarks are classified as fermions, which are a category of subatomic particles with mass. So in addition to quarks, fermions include protons, neutrons, and electrons, but not the photons of force, which are massless.

Quarks do not "fill" protons and neutrons, but are points of infinite mass, making up a small amount of the total mass; the source of the rest of the mass has not yet been determined.

Each **proton** contains three quarks, which determine the *positive* electrical charge:

- ☐ Two "up" quarks; each with $+\frac{2}{3}$ of a positive charge, which total $+1\frac{1}{3}$ positive charge, and one "down" quark, with $-\frac{1}{3}$ negative charge.
- ☐ Combined, the $-\frac{1}{3}$ negative charge cancels out $+\frac{1}{3}$ of the positive charge, leaving the proton with a net positive charge of $+1$.

Compare this to the **neutron**, which also contains three quarks, but which create a neutral charge:

☐ One "up" quark, with +⅔ of a positive charge, and two "down" quarks, each with -⅓ negative charge, which total -⅔ negative charge.

☐ Combined, the -⅔ negative charge of the two "down" quarks exactly cancels out the +⅔ positive charge of the one "up" quark, leaving the neutron with a net neutral charge of 0.

Beta decay. To make things even more interesting, quarks can change from "down" to "up," and from "up" to "down," changing neutrons into protons, and protons into neutrons, in a process called "beta decay".

For example, a single neutron, left alone for 10 minutes in a laboratory, can be observed to "flip" one of its two negatively charged "down" quarks into an "up" quark, resulting in a net positive charge of +1. As part of this process, an electron, with a negative charge of -1, is released and goes into orbit around the newly created proton. At the same time, an ultra-small antineutrino is released. Thus, energy is conserved by neither being created nor destroyed, and balance is maintained.

In the same way, a proton can convert an "up" to a "down" quark to become a neutron. But instead of emitting an electron and an antineutrino, an antielectron, and a tiny neutrino are emitted.

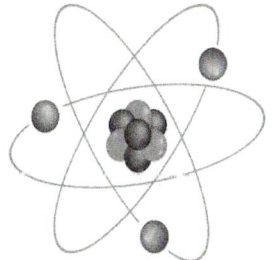

A lithium atom with 3 protons and 3 neutrons.:

Each proton contains 2 up quarks with +2/3 charge and 1 down quark with a -1/3 charge

Each neutron contains 1 up quark with +2/3 charge and 2 down quarks with a -1/3 charge

Before leaving quarks, there are actually six different quarks, all with whimsical names. In addition to "up" and "down," there are "charm," "strange," "top" and "bottom" quarks, but only the "up" and "down" are known to be components of matter; the others were discovered in experiments, but roles for them has not yet been determined.

The Standard Model of Forces: Bosons

Are you wondering why the positively charged protons in the nucleus don't push away from each other, like two positive ends of magnets? Or how protons or neutrons can break free to be emitted during radiation? You just read that neutrons can turn into protons, but what forces are at work? What's more, these particles are not alone within the atom: Three of the four known forces of the universe keep the atom together and enable it to emit radiation. AND to decay into different forms. We'll explore each of these forces.

The Standard Model is what physicists use to describe the unification of the two theories that define all of the subatomic interactions, except for interactions attributed to the force of gravity. In other words, these two theories encompass three of the four known forces of the universe.

Just as the particles with mass are called "fermions", the massless forces are called "bosons".

The first of these is the **electroweak theory**, and it concerns the various interactions that involve two of the four fundamental forces: Electromagnetism, and the weak force, which governs various forms of atomic decay. The second theory is called **quantum chromodynamics,** and it concerns the strong nuclear force.

Not to complicate things, but there's an additional set of theories, called "gauge field theories", and these describe particle interactions involving mediating, or messenger particles, which have a single unit of spin, also called "angular momentum".

The concept of spin can get very complicated and mathematical, so just for the overview, as quantum mechanics began to take hold during the 1920s, Danish physicist Niels Bohr determined that electrons orbit around the nucleus of the atom at fixed angles and velocities, rather than randomly.

Strong force. The most powerful of the four known forces, or bosons, the strong force acts powerfully to hold subatomic particles together. It binds the three quarks in every proton and neutron, and binds the protons and neutrons to hold the nucleus of the atom together. Without this force, the electromagnetic energy within the positively charged protons would repel and force them apart. Despite its intensity, the strong force (also referred to as a particle—the **gluon**), operates over a very short distance.

Weak force. Despite its name, the weak force is second only to the strong force in its strength. It mediates (or facilitates) the conversion of protons and neutrons in beta decay, and is described as a particle, called Z or W, depending on the situation, such as

35

when it changes a neutron to a proton, or when it converts a proton to neutron.

Among other subatomic forces and particles, there are two mysterious yet incredibly important ones: *Dark matter*, and *dark energy*, which we'll cover in the following chapter.

Chapter 4: Neutrinos and the Dark Sector

We began our book with a big step back in time and space to the origin of our universe, 13.8 billion years ago, when everything burst forth from a singular, almost infinitely small point, to the vastness of today's universe, with an estimated hundreds of billions of galaxies, each containing billions of stars, and presumably, billions or trillions of planets. Added to this are black holes, some with the mass of millions of stars, and quasars, whose light is reaching us from 10 billion light years distant (that's 6 trillion miles—the distance light travels in a year— times 10 billion).

Dark Matter and Dark Energy

Yet, this unimaginable amount of mass and energy, what's called the *visible universe,* is now considered to account for only about 5% of what's "out there." Two mysterious entities are believed to represent the remaining 95%: *Dark matter* is being credited with influencing the gravitational effects of the universe, and *dark energy* is believed to be responsible for the unexplainable (and illogical) accelerated expansion of the universe.

In the 1998 NASA reports, the Hubble Space Telescope, in studying distant supernovae (exploded stars), provided data to confirm a pattern of earlier slowing, and more recent expansion. As you read in the Introduction, as the components of the universe expanded outward, after about 6 billion years of gradually slowing due to normal gravitational influences, the expansion of the universe began to accelerate—a phenomenon that defies gravitational forces, and that continues today.

Lacking any other explanation, the acceleration is attributed to an unidentified force, now called "dark energy". Its influence is calculated to be the equivalent of 68% of the mass energy in the universe (recall that Albert Einstein taught us that mass and energy are interchangeable).

Separately, another mysterious entity is asserting a powerful gravitational force that holds stars and galaxies in their tightly constrained orbits. Stars at the outer edge of a spinning galaxy should rotate slower than stars closer to the center, but all the stars rotate at the same speed, regardless of their location within the galaxy. The assumption is that something massive is surrounding the galaxy, pulling all the stars along.

This force is attributed to invisible mass, which is called "dark matter". Based on the effects that can be measured, astrophysicists have calculated that dark matter accounts for 27% of all the matter in the universe, even though it has not been observed directly: Dark matter does not interact with electromagnetic radiation, including light.

Combined, the two "dark" sector entities are believed to account for **over 95%** of the mass of the known universe, and are believed, or at least suspected, to be caused by, or influenced by, yet undiscovered subatomic particles.

Is dark energy causing galaxies to expand?
Is dark matter holding the galaxies together?

Enter the Neutrino

Ever since they were discovered in the 1950s, neutrinos have fascinated and frustrated researchers' efforts to detect and measure them. Here are some fast facts about neutrinos, whose name means "little neutrons," and like neutrons, have no electrical charge.

- [ ] Neutrinos permeate the universe; right now, trillions are passing through you every second at speeds close to the speed of light.
- [ ] They are believed to be created within stars, during the process of nuclear fusion, when hydrogen is fused into helium and heavier elements to generate heat energy.
- [ ] Once thought to be massless, like photons of light, neutrinos are now known to have a tiny, almost insignificant amount of mass. (The fact that neutrinos can't quite achieve the speed of light confirms they are not massless, according to the Standard Model.)
- [ ] But this extremely small mass, along with their neutrality, makes them very hard to detect by the electromagnetic, weak, and strong forces.
- [ ] Adding to their unusual character, neutrinos

change in size during flight from their normal state, called "electron neutrino", to larger "muon", and ever larger, "tau neutrino". But even the tau size is extremely small.

The Sterile Neutrino

So if these three "flavors" or types of neutrinos are so small and undetectable, how can they possibly account for the missing 95% of mass and energy? One possible explanation is their sheer number: Neutrinos are by far the most numerous entity in the universe, more than all the protons, neutrons, and electrons in all the atoms. But even this vast "known" number of neutrinos may greatly understate their true quantity.

The newest hypothesis, as reported by Los Alamos National Laboratory physicists William Charles Lewis and Richard Van De Water in "Scientific American" (2020, July), is the existence of a fourth "sterile" neutrino, that is not just hard to detect; it's virtually *impossible* to detect. Their research, and that of others, implies that the number of neutrinos emitted by our Sun is far fewer than projected; is this because current detectors are missing most of what's coming at us?

So, if the sterile, undetectable neutrinos, the "ghostliest" subatomic particles that Lewis and Van de Water hypothesize, are in the vast quantity they suspect, they could be capable of *interacting* with the dark sector, or even *being dark matter and affecting dark energy*. Confirming this theory is going to be extremely tough, because if it's hard for the first three flavors of neutrinos to be detected by the electromagnetic, weak, and strong forces, it will be impossible for the fourth flavor, the sterile neutrino to be detected.

New technologies are being developed now to be able to track down and assess the elusive fourth flavor. But for now, this is the most current state of the neutrino/dark sector quest for answers.

Next, let's head back to the 1920s and 1930s, when the most brilliant minds came together for a historical set of meetings to confront the irrational, evolving new laws of quantum mechanics.

Chapter 5: The Bohr-Einstein Debates; the Copenhagen Interpretation

Does light travel through space in the form of an oscillating wave, or is it a particle? What is an electron? Uncertainty. Complementarity. Wave-particle duality.

These were key elements of the Bohr-Einstein debates, that are high points of scientific thinking early in the twentieth century. Albert Einstein and Niels Bohr, and their colleagues, including Werner Heisenberg and Wolfgang Pauli, debated concepts, hypotheses, ideas, theories and challenges to scientific knowledge and understanding of the rapidly-evolving discoveries of the atomic and subatomic world. This was the cradle of thinking that established the complex, illogical science of quantum mechanics.

A Positive Relationship

Albert Einstein and Niels Bohr were friends and respected colleagues, and were drawn together by the challenges they shared in defining the hard-to-grasp complexities of the quantum world. One of their disagreements concerned photons: Do they exist?

The 20th century began with physicists in Europe confronting Max Planck's revolutionary concept of the quantum nature of energy, that it does not flow or travel continuously, but rather in discrete packets, or quanta. In 1905, Einstein stepped forward to endorse and enlarge upon Planck's thesis, by proposing that light travels, under certain conditions, as a particle instead of a wave; he called the particle a "light quantum", which later became known as a photon. This was the basis for wave-particle duality.

At this time, Niels Bohr in Copenhagen was having none of this, and would oppose the concept of photons until 1925. Einstein was comfortable with the concept of a photon because he could understand its physical nature, while Bohr was not open to multiple equations for the same phenomenon. In 1913, it was Bohr's turn to theorize, with his hydrogen atom model, a rationale for the atomic spectrum.

It seems simple and obvious to us today, but this was the first time the "solar system" model of an atom was conceived, albeit with electrostatic force controlling the orbiting electron, rather than gravity affecting orbiting planets. Einstein was initially opposed to the concept, but later supported it, which is described as follows:

- ☐ The electron orbits around the nucleus of the atom, but does not radiate or emit energy; this is in opposition to earlier classical physics.
- ☐ The orbit of the electron is only at a discrete distance from the nucleus (not in between); these specific orbits are now called "energy shells".
- ☐ There is a point at which an electron can get no closer to the nucleus. Electrons can gain or lose energy when they jump to a higher level orbit shell, or descend to a lower one.

The Uncertainty Principle

One of the most vocal and influential physicists at the time was Germany's **Werner Heisenberg**, who is best known for his uncertainty principle of 1927, which exemplifies the vague, indeterminate nature of quantum mechanics. The principle posits that the more that is known about the position of a subatomic particle, the less can be known about its momentum. Similarly, the more you know about a particle's momentum, the less accurate can be your reading of its position. Why is this?

Heisenberg (and others) traced the rationale to the *matter wave* nature of all quantum entities and objects. The hypothesis of the behavior of matter as a wave was developed in 1924 by **Louis de Broglie**, and it supports the concept of wave-particle duality. Matter waves are known as *de Broglie waves*.

Niels Bohr with Albert Einstein
December 1925

Niels Bohr
1922

Werner Heisenberg
1933

The uncertainty principle should not be confused with the similar *observer effect*, which says that measurements of subatomic particles cannot be taken without disrupting or changing a component of the system. For example, the energy needed to measure the position of an electron or a photon will disrupt the particle's properties at that moment.

Einstein's Argument Against Bohr's Complementarity

During the Fifth Solvay International Conference in 1927, which concerned electrons and photons, Albert Einstein opposed the existing orthodoxy. This included the well-accepted first law of thermodynamics, known as the law of conservation of energy, plus the law of momentum, to gather information on the state of a particle while undergoing interference. Einstein claimed that his position, based on those two laws, refuted Niels Bohr's popular principle of *complementarity*.

Bohr's complementarity principle appears similar to Heisenberg's uncertainty: Objects have pairs of properties that are complementary, which prevents the properties from being measured or observed at the same time—that is, simultaneously.

But according to Einstein's proof of concept, when a monochromatic light beam meets a screen, it is diffracted, and continues on to a second screen with a pair of slits. This results in an interference pattern (F, in the schematic). Assuming only one particle can pass through at a time, and having the ability to detect the impact of the particle on the screen, it is possible to calculate which slit the particle passed through without disrupting the wave quality of the particle.

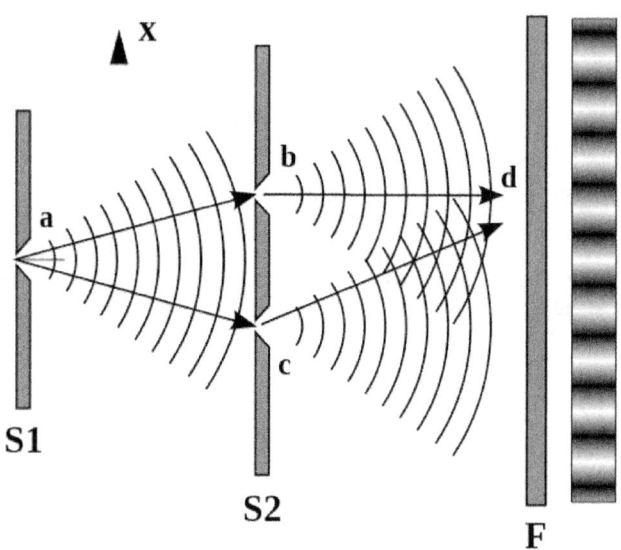

If there was a single overriding conclusion to the Einstein-Bohr debates, it is that neither of them was completely confident in their summaries of exactly what quantum mechanics is really about. Yet, both held to their views, believing they were close to reality, if not completely correct. Physicists, then and now, concur that Bohr was successful in his defense of quantum theory, and in establishing the probability character of quantum measurement:

> "Bohr was inconsistent, unclear, willfully obscure and *right*. Einstein was consistent, clear, down-to-earth and *wrong*." (John Stewart Bell, as cited by Graham Farmelo, The New York Times, 2010, June 11).

The Copenhagen Interpretation

Bohr's Copenhagen interpretation, in collaboration with Werner Heisenberg, is still a widely-accepted description of quantum mechanics. Between 1925 and 1927, Werner Heisenberg was frequently invited to Bohr's home in Copenhagen, Denmark, for meetings at home, and on walks in the forest, to debate the meaning of the science of quantum mechanics. Their collaboration brought forth one of the earliest interpretations, yet is still popularly taught today.

Basically, for the specific, or definite properties of physical systems, until they are measured, and even when measured, the findings are conditional, or probaliatic; that is, quantum mechanics can only predict a *probability* distribution of results. It is reminiscent of Heisenberg's uncertainty principle, and Bohr's complementarity explanation.

Consider this to be like the odds in a horse race, with some horses more likely to win than others, but no definitive prediction of the race outcome is possible. Once a measurement is made, (just as when the horse race is over and one horse is the winner), one, and only one definite result is possible. Physicists call this effect "wave motion collapse," which is the reduction of a large number of possibilities to a single state.

Summing up, most physicists agree that the Copenhagen interpretation is largely regarded as the same as numerous theories: Indeterminism, Bohr's

Correspondence Principle, his wave function statistical interpretation, and his complementarity conception of various atomic phenomena.

Moving forward in the following chapter to more contemporary times, another debate wrestles with an ambitious concept; that matter is ultimately composed of infinitely small, vibrating strings.

Chapter 6: String Theory: A Unified Theory?

String theory is a popular, contemporary concept that was developed to make the Standard Model complete by unifying quantum mechanics with general relativity.

Einstein's 1915 general theory of relativity redefined gravity as the warping and bending of space and time, or space-time. As brilliant as this discovery was, it failed to unify gravity with the three other known quantum forces of electromagnetism, the weak force, and the strong force. Einstein's goal, and that of physicists ever since, was to combine quantum mechanics and general relativity in a single quantum mechanical perspective—the unified theory.

What is String Theory?

It proposes that protons, neutrons, and the quarks they contain are not the smallest entities that comprise matter. Rather, there are infinitely smaller, string-shaped entities, instead of more traditional concepts of subatomic point particles with zero dimensions. These strings are proposed to vibrate relative to the charge and mass of particles, and to actually be the source of those properties. Initially, strings were thought to be one-dimensional, but that idea has changed dramatically, as you will see.

The search for unification: Since the 1920s, theoretical physicists have tried to accurately describe the processes of atoms, subatomic particles and forces within the evolving science of quantum mechanics. But a continuing problem with general relativity comes in: The masses of particles at the quantum level are so small that gravity essentially has no effect upon them. To put this in perspective, gravity is estimated to be some billion, billion times weaker than any of the other three forces. So when it comes to subatomic particles, gravity is simply a "no show."

Gabriele Veneziano: The Strong Force

With gravity off the table, by the late 1960s, as the other three forces were being studied, the strong nuclear force, which holds together the protons and neutrons in the nucleus of atoms, was gaining the most attention.

In 1968, Gabriele Veneziano, a young physicist working at CERN, the European Center for Nuclear Research, came to the realization that the data being collected by particle colliders was leading to new conclusions regarding the nature of the strong force.

Within three years, physicists were amplifying Veneziano's findings: Stamford's Leonard Susskind, Niels Bohr Institute's Holger Nielsen, and University of Chicago's Yoichiro Nambu, interpreted the underlying mathematics to define miniscule energy filaments in vibrational motion; these filaments are like tiny bits of string. Thus, the inspiration to call this concept of the strong force, "string theory" came about.

But then the going got rough for string theory a few years later, when new data on the strong force challenged the validity of the supportive mathematics.

Many theoretical physicists turned away from string theory, but then scientists at the California Institute of Technology in Los Angeles, the École Normale Supérieure in Paris, and Japan's Hokkaido University, came to the conclusion that a *supposedly failed* prediction of string theory—that the strong force was based on a massless particle, was actually proved by the inability of the experiments to find a particle with mass. They thought that this finding might finally be proof of Einstein's forecast unification of theories.

Since its inception, quantum theory and relativity theory have been in opposition and incompatible. General relativity's laws of gravity did not fit with the other three forces, but now, the possible existence of a massless particle that string theory predicted could bring unification. This would take string theory far beyond a description of the strong force...

Extra Dimensions

But as usual, in all unification theories, there were problems, most notably the requirement that the strings be multidimensional. All matter as we know it in both classical and quantum physics have three dimensions, commonly called: Height, width (or length), and depth. But the math of string theory accords six more dimensions, totaling nine, or ten, if the space-time dimension is counted.

The problem in detecting these strings is the assumption that the extra six dimensions are crumpled, making them even smaller and harder to detect than originally thought. Being massless further adds to the difficulty in detecting them.

But new interpretations of string theory continued to evolve. An analogy to a violin string's vibration patterns playing different notes was used to suggest that strings vibrate differently for each particle in nature; one vibration for an electron, another vibration pattern for a quark.

This theory states that strings are so small that they are points without length, but more recent calculations predict a length, albeit one that is incredibly small, even by subatomic scales. Significantly, it was believed that strings also vibrate uniquely for each of the four forces, and if this were confirmed, it would be the long-sought unification of all matter and forces in the universe.

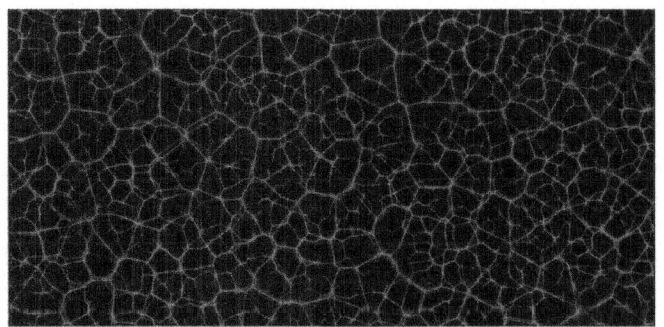

Are infinitesimally tiny, vibrating, multidimensional strings
the foundation of all matter?

Then in 1995, Edward Witten of the Institute for Advanced Study refined many of the existing equations and discovered that strings may have a seventh unique dimension, bringing the total up to 11: Seven new dimensions, the three existing, plus space-time. Witten also found that the theories revealed a membrane-like substance co-existing with strings; these became known as "brames".

Witten's work gave string theory a new momentum, and the refined set of equations became known as **M-theory** (with no indication of what "M" stands for—membranes, possibly).

Supersymmetry

Within string theory is the concept of supersymmetry, in which every type of particle has a "superpartner," and this is why string theory is also called *superstring* theory. While no superpartner has yet to be detected, search is underway at the giant particle accelerator at CERN, on the French-Swiss border. Superpartners are estimated to be larger than their original partners. If discovered experimentally, supersymmetry will add circumstantial support to string theory, and add to the possibility that the four forces can be unified.

Symmetry implies that a particle or other entity can undergo transformation, yet remain unchanged. A common demonstration of this effect is when a square is rotated, and appears the same in each of its four positions, from zero degrees through, 90, 180, 270 and finally back to 360 degrees in the starting position.

Supersymmetry allows fermions (like quarks and electrons) to convert to bosons (like the strong and weak forces) without undergoing any change in their fundamental theoretical structures. This conforms with the laws of conservation of energy and of momentum, and reduces the number of basic particles from two to one, by showing they are essentially interchangeable. Supersymmetry extends from string theory to the quantum field theory of gravitational forces in unifying the four forces in the Standard Model of quantum physics.

As of today, string theory remains an encouraging yet challenging concept, with no experimental proof of its validity (so far). It operates at a level so infinitesimally tiny that it cannot yet be observed with current technology. Testing continues, theories are revised, and the hunt to confirm string theory continues.

Continuing to explore the mysterious world of quantum physics, let's head on to an understanding of the unusual ideas of Schrödinger's cat.

Chapter 7: Schrödinger's Cat and Superposition

When we think of Edwin Schrödinger's 1935 thought experiment involving the well-being or demise of a cat, let's insert a motion-picture-type of disclaimer: *No animal was harmed in the making of this example*. It's not an experiment that was ever performed; the celebrated cat is fictional, and its fate is conditional. It's all about the concept of uncertainty.

A Demonstration of Uncertainty

There is a duality in quantum mechanics, based on uncertainty, which Schrödinger thought to demonstrate with a whimsical example. First, let's consider the inability to know when a radioactive atom will decay. The individual atoms in a radioactive element, like radium, decay with a frequency that cannot be precisely determined. The half-life of the element tells you whether half its atoms will decay within an hour, a day, a year, or 100 million years, but there's no way to know exactly when one of those atoms "decides" it's time to fire off an alpha particle, which is a two-proton, two-neutron helium atom nucleus. It's uncertain.

Schrödinger's thought experiment was conceived to demonstrate this uncertainty in the subatomic world by combining it with an example of uncertainty in a real-world, macroscopic environment.

A box is brought forth, and a cat is placed inside the box. Now a radioactive element is placed in the box, along with a flask containing a highly toxic poison; the flask is positioned so if the element decays and fires off an alpha particle—it will break the flask, release the poison, and (hypothetically) kill the cat. The box is then sealed for a brief period.

According to quantum theory, the radioactive element either has, or has not, decayed, and the cat is, or is not, alive. As long as the box is sealed, and nothing inside can be observed, the cat is said to be alive and dead at the same time. Only when the box is opened can the observer know, with complete certainty, the state of the cat's well-being.

The Schrödinger Equation

Schrödinger was making a fanciful, extreme case to demonstrate uncertainty. Over time, his cat-in-a-box example took on a life of its own, and became a popular, if an overly-simplistic way of describing uncertainty. He was not advocating the idea that a cat could really be both dead and alive at the same time. He was being contemptuous of what he thought was the absurdity of contemporary interpretations of quantum science.

Who exactly was Edwin Schrödinger? An Austrian physicist and contemporary of Einstein, de Broglie and other pioneers of quantum physics, Schrödinger gained prominence by defining the wave function (state) of a quantum mechanical system or configuration. The Schrödinger Equation was published in 1926, and became a foundation of quantum theory. Schrödinger received the 1933 Nobel Prize in Physics for this work.

According to the **Copenhagen interpretation** of quantum physics (see chapter 5), the wave function, as defined by Schrödinger, is the most comprehensive definition of all physical systems. This includes not only at the atomic, molecular, and subatomic levels, but also macroscopic systems. Some interpretations include all functions within the universe.

Closer to our time, as new ideas and interpretations of quantum science have come forward, the concept of superposition, with the cat both dead and alive at the same time remains relevant, and is still being used to demonstrate the qualities and shortcomings of various new interpretations.

Quantum Superposition

Can more than one reality exist, at least at the quantum level? According to quantum mechanics, a photon that is not measured is in a state of superposition—meaning that it exists in multiple states at the same time. It is only when it is measured that the photon's position, or its momentum, becomes fixed and singular.

Quantum superposition is a variation of the concept underlying Schrödinger's thought experiment, which contends that atoms and photons of energy can be in different states, depending on various potential outcomes. According to the Copenhagen interpretation, the particles and forces remain in a state of superposition, meaning that their actual state is uncertain and cannot be predicted until observed.

Einstein, at about this same time, was publishing a perspective that the superposition collapses into one or another of the possible definite states. He contended that an experiment can demonstrate that a system with particles that are apart by considerable distances may exhibit the qualities associated with a state of superposition. Einstein used the example of a barrel of gunpowder, saying it could contain a superposition of both unexploded and exploded positions—again, with that state dependent on an observation to make it definitive. This example is thought to have inspired Schrödinger's famous thought experiment.

Superposition contends that multiple realities can exist at the quantum level. According to quantum mechanics, a photon that is not measured remains in a state of superposition; it exists in multiple states at the same time. It is only when it is measured that the photon's position becomes fixed and singular—that is, the photon "chooses" a single state or position.

Physicists have recognized that superposition has only been observed at the quantum level, such as the wave particle duality demonstrated by the double slit experiment (see chapter 2). A team led by Markus Arndt of the University of Vienna developed an experiment to see how a large particle can be proven to exhibit the same superposition as the singular photons and waves used in the original experiments.

Using a high-tech instrument called an "interferometer", the team used a nanosecond laser pulse of light photons to fire large molecules towards slotted screens, with the objective of creating patterns on a back screen. Molecules as large as 2,000 atoms performed like waves, even though they were particles.

Another aspect of the strange behavior of particles at the quantum level is entanglement; coming up next. Both entanglement and superposition were found to have practical applications, especially quantum computing.

Chapter 8: Entanglement and Quantum Computing

In simplest terms, entanglement is a link or connection between distant particles, so when one particle is changed or disturbed, it affects the other particle instantly. It does not matter if the particles are close or far apart.

Spooky Action at a Distance

Einstein called entanglement "spooky action at a distance." The unusual behavior of particles linking or interacting over a distance defies current theories, including limitations imposed by the speed of light, yet experiments are leading to practical applications, including quantum computing. Entanglement is a difficult concept to grasp, even in the unpredictable world of quantum physics, where uncertainty prevails over predictability. But we can break it down into its basics.

The concept of entanglement as a function of quantum mechanics has been known and debated for decades, but now, technology is testing and proving that entanglement is real.

When Einstein subsequently made his "spooky action at a distance" remark, he was questioning the logic of their findings, and asking, in effect: If the currently-believed conceptualization of quantum physics is correct, then entanglement should not be possible. Ergo, if entanglement truly is possible, then our quantum theories are incomplete. On a more technical level, entanglement violates the local realism view of causality. This led to what has become known as the "EPR paradox".

The EPR Paradox

As early as 1935, a paper entitled "Can the Quantum Mechanical Description of Physical Reality be Considered Complete?" was collaborated on by Albert Einstein, Boris Podolsky, and Nathan Rosen. This paper, referred to as the "EPR paradox", established the potential for entanglement, even though these same theoretical physicists were dubious about it, and questioned its feasibility. Edwin Schrödinger also wrote papers describing the phenomenon.

The EPR paper presented a case for elements or components of a state of reality that somehow exists outside current quantum physics. Therefore, a new broader theory that would include entanglement would be needed. In resolving the EPR paradox, important improvements to quantum mechanical theory would emerge.

A thought experiment was developed, involving two particles in an entangled state. If entanglement was real, when the position, or the momentum, of one of the particles was measured, the finding would make it possible to accurately predict the position, or momentum, of the second particle. Yet, they argued, this would violate the theory of relativity that prevents information from traveling instantaneously—that is, faster than the speed of light.

The three scientists stated a principle that became the "EPR criterion of reality," that the second particle would have already had values of position and momentum (in other words, it is not because information traveled from the first particle faster than light). This conclusion conflicted with the views of Niels Bohr and Werner Heisenberg, who believed that a quantum particle cannot have the specific values of position and momentum until it is measured.

So, what are we to believe? Is entanglement a confirmed physical phenomenon today? Theories and thought experiments aside, can information between distant particles actually travel faster than light? No, according to all current knowledge, yet the joining of particles does appear to exist. Entanglement has been experimentally demonstrated with photons, electrons, neutrinos, and molecules. Entanglement is playing a role in certain forms of communications, quantum computing, and quantum radar.

Examples of recent experiments that are validating

entanglement's existence:

- ☐ In Brazil, Gabriela Barreto Lemos and associates conducted quantum imaging tests in 2014, using photons to photograph objects that had not interacted with the subjects, but were entangled with entangled photons that did interact with them; a technique that may have application in medical imaging.
- ☐ In 2015, a team at Harvard University, led by Markus Greiner, made direct measurements of entanglement in a system of super cold atoms.
- ☐ Beginning in 2016, IBM, Microsoft and other technology leaders have developed quantum computers, and their developers were encouraged to experiment with new quantum entanglement concepts.

What can we conclude? If quantum entanglement is increasingly being confirmed to exist, can it be presumed that quantum mechanics, as currently understood, is not complete? The answer may be neither: Entanglement may exist, but probably not exactly as it has been described, so far. And quantum mechanics remains incomplete, until a unification theory can be proven. In time, we should expect to learn of new definitions of entanglement that are less "spooky," and more compatible with established quantum science.

Quantum Computing

While the unknowns of entanglement continue to provoke debates into its mysteries, quantum mechanics marches along with real-world applications. Entanglement and superposition are being applied to the rapidly-emerging field of quantum information science, including quantum computing. Quantum computers are showing the potential to do more and faster work than traditional computers, despite the blazing speed and gigantic memories of today's servers and supercomputers.

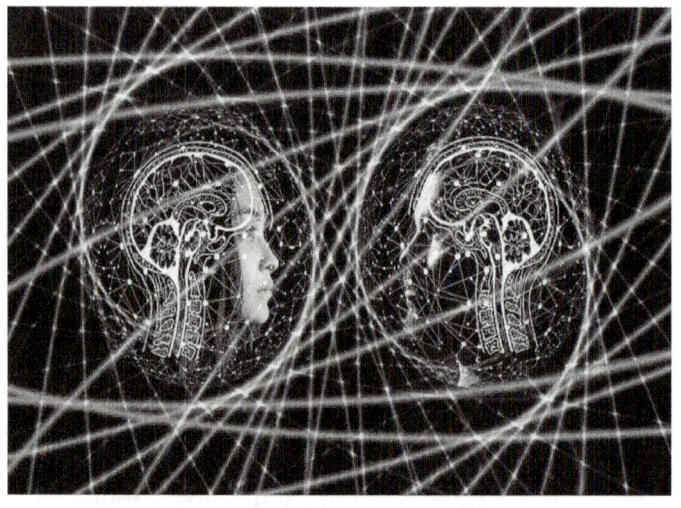

A brief history: Physicist Paul Benioff first conceptualized a quantum model of the Turing machine—an early form of computer that was developed by the UK's Alan Turing during WWII to crack the German Enigma code. Later, Yuri Manin and Richard Feynmann proposed that a quantum computer could have unique computational capabilities, and in 1994, a quantum algorithm for decryption was developed by Peter Shor.

Here's how it works: The classical computers we all use are based on what are called "bits," which are two digits: 1 and 0, representing "on" or "off," and processed by an electrical signal. By stringing together large multiples of these "on" and "off" bits, we can process text, math, music, video, and everything digital.

In contrast, a quantum computer's bits (called qubits) are composed of subatomic particles—usually photons or electrons. As quantum particles, they can have unique quantum properties, including superposition and entanglement. Superposition enables qubits to exist in multiple states, so that a qubit can encode 1 and 0 at the same time. Entanglement allows two qubits to link, and despite a separation, to influence each other. So, changing one qubit instantly changes the other, shortening computing time appreciably.

The qubits in a quantum computer are initially in *both* a 1 and 0 combined state, as the computer begins to work through a problem, but as soon as the solution is developed, the qubits drops from superposition to a single set of 1 and 0, making a finite, specific result possible.

Don't expect your next desktop or laptop computer to be a quantum computer; for us, that may be many years away. But for the most challenging problems, the ability to process multiple possibilities simultaneously may soon be applied to highly complex atomic structural modeling, including economic market forecasts, genetic engineering, and long-range weather forecasts.

Now, let's move from the very small to the very large, with an exploration of parallel universes.

Chapter 9: Parallel Universes: Do They Exist?

As you will recall from the Introduction, our universe has been expanding for 13.8 billion years, and since photons of electromagnetic energy travel at the speed of light in every direction, the universe should be at least 27.6 billion light years across. But astrophysicists project a diameter of 93 billion light years, because space itself is expanding, further separating matter and energy.

Does this mean that objects are moving away from each other faster than the speed of light? Special relativity theory does not allow *local region* objects from exceeding light speed, but according to Hubble's law, there are no constraints when the space separating far distant objects is expanding.

But as awesome as all this is, there may be more—much more.

The Multiverse Concept

While the vast size of the universe seems hard to comprehend, it gets even more mind-boggling to consider theories that imply that "our" universe may be only one of a multitude of universes; there are astronomical theories that explore a multiverse. Some describe a vast diversity, while others explore parallel universes that are virtually identical to each other.

Elizabeth Howell is a Ph.D. candidate in aerospace sciences, who wrote in *Space.com* (2018), that there are *five multiverse options* currently under consideration:

1. **Infinite number:** Since space-time is without a defined shape, it may extend infinitely, leaving open the possibility that matter and energy can duplicate themselves, and form an infinite number of "other" universes, far beyond the range of our telescopes. In other words, our universe's expansion may

eventually catch up with other universes, which might be expanding in our direction as well. Or, the infinite number of universes may forever continue to exist in isolation from each other.

2. **Many bubbles:** Studies conducted at Tufts University by Alexander Vilenkin imply that space-time may be slowing or stopping in some regions, and continuing to expand in others, opening the possibility of this universe being one of many "bubbles" across space. Being separate, each bubble universe could have its own distinct laws of physics and chemistry.

3. **Many daughters:** Based on the laws of uncertainty and quantum mechanics, every situation has multiple options until observed or measured, so possibly our universe is one of many alternatives—similar yet slightly different. We exist in each daughter universe, but with slight variations. As there is no one to "observe" these universes, their positions will remain unknown.

4. **Mathematical:** As seemingly fixed and unchangeable as mathematical laws are, MIT's Max Tegmark theorizes that the structure of mathematics can vary by different universes.

Two plus two may not equal four. This is because mathematics is a human construct, and in universes without humans, totally different concepts of mathematics may exist.

5. **Parallel universes**: This concept is based on flat space-time, which would permit an almost-limitless number of particle configurations. Thus, there are an infinite number of universes that repeat themselves with every possible arrangement, from exactly what our current universe is like (including you and everyone you know) to completely unrelated arrangements. This recalls the definition of "infinity" that says an infinite number of monkeys with an infinite number of typewriters or laptops will eventually recreate a play of Shakespeare.

The Search for Evidence

A 2015 article in *Medium* by astrophysicist Ethan Siegel questioned whether parallel universes, closely similar to each other, would be feasible. He notes that the age of this universe is 13.8 billion years, meaning that because its age is not infinite, an infinite number of nearly-identical parallel universes is highly unlikely. Our own universe appears to be unique, Siegel says.

Could a black hole lead to another universe?

Studies analyzing data from the Wilkinson Microwave Anisotropy Probe in 2010 indicated that our universe collided with a parallel universe at some point in the past, raising expectations that "there's something else out there." But deeper analysis of those data and findings from the higher resolution Planck satellite found no evidence of a bubble-universe collision, and further no evidence of the gravitational attraction such an entity would have created.

While some theories imply an infinite number of parallel universes, Stephen Hawking, in one of his last papers published in 2018, said that there may be more than a single, unique universe, but his research findings suggest limiting the multiverse to a small range of potential universes.

When cosmologist Paul Davies wrote "A Brief History of the Multiverse" in *The New York Times* (2003), he challenged all of the multiverse and parallel universe theories as being non-scientific, and completely without even partial evidence. He acknowledged the limits of our telescopes to reach beyond a certain distance, but said that to believe in multiverses and bubble universes depends completely on faith, making it like a theological "leap of faith." No multiverse concept can stand up to the *scientific method* of evaluation.

A further perspective was offered by theoretical physicist George Ellis. Writing in *Scientific American,* he proposed "an open mind, but not too open." In the absence of any scientific evidence, we will have to live with uncertainty about the multiverse:

> "As skeptical as I am, I think the contemplation of the multiverse is an excellent opportunity to reflect on the nature of science and on the ultimate nature of existence: Why we are here. In looking at this concept, we need an open

mind, though not too open. It is a delicate path to tread. Parallel universes may or may not exist; the case is unproved. We are going to have to live with that uncertainty. Nothing is wrong with scientifically-based philosophical speculation, which is what multiverse proposals are. But we should name it for what it is."

—George Ellis, *Scientific American (2015, December)*, "Does the Multiverse Really Exist?"

The quest for proof of alternative or parallel universes continues, and as telescopes and detection technology become more advanced, the yet-undiscovered evidence may be found. Many scientists believe it is not a matter of "if," but of "when."

Conclusion

The preceding chapter provides a concluding note to our study of quantum mechanics: Conceptual thinking can guide the exploration of the unknown, and lead to new discoveries, but until scientific proof exists, we are left with concepts, theories, and hypotheses. As George Ellis puts it so clearly:

> "Parallel universes may or may not exist; the case is unproved. We are going to have to live with that uncertainty."

The Scientific Method

The *scientific method* is the process by which a theory or hypothesis is tested: It eliminates bias, accepts only facts, and requires reproducibility—that the same results will be obtained by every test. The half-life of radium, whose isotopes each decay to exactly half their current level of radioactivity according to research-confirmed, predictable timelines:

- [ ] Radium 223: 11.4 days
- [ ] Radium 224: 3.64 days
- [ ] Radium 226: 1,600 days
- [ ] Radium 228: 5.75 years

Meanwhile, creative minds are exploring the yet-unresolved mysteries of quantum mechanics, facing the uphill battle of trying to more completely understand the infinitesimally small world of subatomic particles and forces. Let's conclude our discussion with a relevant and contemporary example—the Higgs boson.

The Higgs Boson

The search for the elusive Higgs boson has finally revealed the particle believed to be responsible for giving mass to matter, and ending one of the great mysteries of quantum mechanics. It took over four decades and the world's largest particle accelerator to bring the Higgs boson out of hiding. Why is the Higgs boson so important?

The Higgs boson was theorized in 1964 by University of Edinburgh physicist Peter Higgs and others as a massive carrier particle that permeates space throughout the universe, and interacts with subatomic elementary particles to give them mass. If confirmed, the Higgs boson would provide a complete understanding of the structure and nature of matter. Importantly, this was considered to be a testable hypothesis, which triggered decades of experimentation.

Briefly, the different mass levels of each subatomic particle is attributed to the particle's unique strength of interaction with the Higgs field. It explains why photons (carriers of electromagnetic forces) have zero mass, while the weak force carrier particles have considerable mass. Finding the Higgs particle would confirm the existence of the Higgs field.

The experiment to find the massive Higgs boson involved the world's largest high-energy particle colliders: Tevatron and the Large Hadron Collider (LHC) in France and Switzerland. The Higgs boson was finally identified at LHC in 2012, and confirmation came in March 2013. Later that year, Higgs and a Belgian associate were awarded the Nobel Prize in Physics.

A particle accelerator where protons are collided at close to the speed of light, to investigate the structure of matter.

In coming to the end of this adventure into the mysteries and discoveries of quantum physics/quantum mechanics, I hope this has provided the illumination you expected. If so, please leave a positive rating, to encourage others to share in this knowledge.

Warmest regards,

Donald B. Grey

Citations

Andal, J. (2014, January 17). Special relativity simplified.
    *Futurism.*https://futurism.com/special-relativity-simplified

Atkinson, N. (2014, February 18). Quantum entanglement explained. *Universe*
    *Today.*https://www.universetoday.com/109525/quantum-
    entanglement-explained/

Boyd, A. (2019). The Bohr-Einstein debates. *University of Houston*
    *Engineering.*https://www.uh.edu/engines/epi2627.htm

Bray, B. (2019, November 25). How do you explain quantum entanglement to
    someone in layman's terms? *Quora.* https://www.quora.com/How-
    do-you-explain-quantum-entanglement-to-someone-in-laymans-
    terms?share=1

Choi, C. (2017, June 17). Our expanding universe: age, history, and other facts.
    *Space.*https://www.space.com/52-the-expanding-universe-from-the-
    big-bang-to-today.html

Davies, P. (2003, April 12). A brief history of the multiverse. *The New York*
    *Times.*https://www.nytimes.com/2003/04/12/opinion/a-brief-
    history-of-the-multiverse.html

Einstein, A.,Infeld, L. (1938). The Evolution of Physics: the growth of ideas
    from early concepts to relativity and quanta. *Cambridge University*
    *Press.*Bibcode:1938epgi.book.....

Ellis, G. (2015, December). Does the multiverse really exist? *Scientific*
    *American.*https://www.scientificamerican.com/article/does-the-
    multiverse-really-exist1/

Encyclopedia Britannica. (2020). Subatomic particles.
    https://www.britannica.com/science/subatomic-
    particle/Elementary-particles#ref367641

Everything Explained Today. (2020). Quantum entanglement explained.
http://everything.explained.today/Quantum_entanglement/

Everything Explained Today. (2020). EPR paradox explained.
http://everything.explained.today/EPR_paradox/

Farmelo, G. (2010, June 11). Random acts of science. *The New York Times*.https://www.nytimes.com/2010/06/13/books/review/Farmelo-t.html

Greene, B. (2017, December 1). String theory. *Encyclopedia Britannica*.https://www.britannica.com/science/string-theory

Hamar, A. (2019, August 1). The double-slit experiment cracked reality wide open. *Discovery*.https://www.discovery.com/science/Double-Slit-Experiment

Hendricks, J. (2015). Quantum physics: Superstrings, Einstein and Bohr. *Amazon*.https://www.amazon.com/Quantum-Physics-Superstrings-Electrodynamics-Dimensions-ebook/dp/B00WPFOYKA/ref=sr_1_3?s=digital-text&ie=UTF8&qid=1549701779&sr=1-3&keywords=Quantum+Physics+For+Beginners#reader_B00WPFOYKA

Hosch, W. (2020). Compton Effect. *Encyclopedia Britannica*.
https://www.britannica.com/science/Compton-effect

Howell, E. (2018, May 10). Parallel universes: theories and evidence. *Space*.https://www.space.com/32728-parallel-universes.html andhttps://videos.space.com/m/qpwLmOG0/in-quantum-physics-more-than-one-reality-exists?list=9wzCTV4g

Howell, E. (2017, November 7). What is the Big Bang Theory? *Space*.https://www.space.com/25126-big-bang-theory.html

Howell, E. (2017, March 17). Einstein's theory of Special Relativity. *Space*.https://www.space.com/36273-theory-special-relativity.html

Hoyer, S. (2011, March 10). What is the difference between quantum physics and quantum mechanics? *Quora*.https://www.quora.com/What-is-

difference-between-quantum-physics-and-quantum-
mechanics?share=1

Leman, J. (2019, October 2). A quantum leap in the classical world. *Popular Mechanics*.https://www.popularmechanics.com/science/math/a29339863/quantum-superposition-molecules/

NASA. (2020). Dark matter, dark energy. *NASA Science*.https://science.nasa.gov/astrophysics/focus-areas/what-is-dark-energy

National Geographic. (2020). Dark matter and dark energy. https://www.nationalgeographic.com/science/space/dark-matter/

Physics of the Universe. (2020). Important scientists: Max Planck. https://www.physicsoftheuniverse.com/scientists_planck.html

Space. (2020). In quantum physics, more than one reality exists. *Space*https://videos.space.com/m/qpwLmOG0/in-quantum-physics-more-than-one-reality-exists?list=9wzCTV4g

Stanford University. (2019, December 6). Copenhagen interpretation of quantum mechanics. *Stanford Encyclopedia of Philosophy*. https://plato.stanford.edu/entries/qm-copenhagen/

Sutton, C. (2020). Higgs boson. *Encyclopedia Britannica*.https://www.britannica.com/science/Higgs-boson

Sutton, C. (2020). Unified field theory. *Encyclopedia Britannica*.https://www.britannica.com/science/unified-field-theory

Sutton, C. (2020). Strong Force. *Encyclopedia Britannica*. https://www.britannica.com/science/strong-force

Sutton, C. (2020). Supersymmetry. *Encyclopedia Britannica*. https://www.britannica.com/science/supersymmetry

Terrasi, J. (2019, October 1). What is quantum computing? https://www.digitaltrends.com/computing/what-is-quantum-computing/

Waldrop, M. (2017, May 16). Einstein's relativity explained in 4 simple steps. *National Geographic*.https://www.nationalgeographic.com/news/2017/05/einstein-relativity-thought-experiment-train-lightning-genius/

Wikipedia. (2020). Complementarity (physics). https://en.wikipedia.org/wiki/Complementarity_(physics)

Wikipedia (2020). Einstein-Bohr debates. https://en.wikipedia.org/wiki/Bohr%E2%80%93Einstein_debates

Wikipedia (2020). Multiverse. https://en.wikipedia.org/wiki/Multiverse

Wikipedia (2020). Photoelectric effect. https://en.wikipedia.org/wiki/Photoelectric_effect

Wikipedia. (2020, April 20). Schrödinger's cat. *Simple English Wikipedia*.https://simple.wikipedia.org/wiki/Schr%C3%B6dinger%27s_cat

Wikipedia. (2020). Schrödinger's equation. https://en.wikipedia.org/wiki/Schrödinger_equation

Wikipedia. (2020). Wave function collapse. https://en.wikipedia.org/wiki/Wave_function_collapse

Wikipedia. (2017, September 18). Quantum entanglement. *Simple English Wikipedia*.https://simple.wikipedia.org/wiki/Quantum_entanglement

Wikiquote. (2019, September 13). Bohr-Einstein debates. https://en.wikiquote.org/wiki/Bohr%E2%80%93Einstein_debates

Zimmerman Jones, A. (2018, March 26). What is the Compton Effect and how it works in physics. *Thought Co*. https://www.thoughtco.com/the-compton-effect-in-physics-2699350

Image Sources

All historical black-and-white images of physicists are in the Public Domain and have been sourced from www.wikipedia.com

All other images have been sourced from Pixabay. https://pixabay.com or Unsplash. https://unsplash.com

Bluesource And Friends

This book is brought to you by Bluesource And Friends, a happy book publishing company.

Our motto is **"Happiness Within Pages"**

We promise to deliver amazing value to readers with our books.

We also appreciate honest book reviews from our readers.

Connect with us on our Facebook page www.facebook.com/bluesourceandfriends and stay tuned to our latest book promotions and free giveaways.